Praise for
Do Unto Earth

"A must-read book for everyone who cares about the future of humanity and our planet."

—**Dr. Ervin Laszlo**, two-time Nobel Peace Prize nominee, recipient of the Goi Peace Prize and International Mandir of Peace Prize, best-selling author of Science and the Akashic Field, founder of the Laszlo Institute of New Paradigm Research and The Club of Budapest, fellow of the World Academy of Art and Science and the International Academy of Philosophy of Science

"A 911 call from Planet Earth herself, *Do Unto Earth* is a potent manifesto for living life today and forward. This book should be required reading in schools. We must act now!"

—**Mary Madeiras**, three-time Emmy-Winning director, screenwriter, Akashic Records practitioner, activist, and author

"*Do Unto Earth* is full of empowering messages and mind-bending assertions that you won't find in science or history textbooks. Given the urgent need for new solutions on this endangered planet, the ideas are worthy of further investigation."

—**Mark Gober**, author of *An End to Upside Down Thinking*, board of directors of the Institute of Noetic Sciences (IONS) and the School of Wholeness and Enlightenment (SoWE)

"From page one, I was hooked! *Do Unto Earth* merges spirituality with our environmental crisis and does it in a way that is as gripping as a blockbuster movie. Brava to Hayes, Borgens…and Pax."

—**Temple Hayes**, author, spiritual leader, animal activist, and founder of illli.org

"The channeled Spirit energy Pax states that we are at the 'crossroads of our survival' and offers us bold envisioning and direction. Mother Earth is speaking, and ancient mysteries are revealed! Let's heed and implement these game-changers for the benefit of us all."

—**Sunny Chayes**, social/human rights and environmental activist, feature writer and Chief Strategic Partner for Whole Life Times, and host of ABC's *Solutionary Sundays*

"Timely, high-level and generative wisdom detailing how we may still sustain our beautiful planet while reclaiming our collective and individual sovereignty."

—**Stephan McGuire**, director of Zürich-based NGO Cernunnos Media, Director of Tree Media Foundation

Pax and the Next

Evolutionary Leap

Pax and the Next Evolutionary Leap

Volume 7 of Do Unto Earth

PENELOPE JEAN HAYES,
CAROLE SERENE BORGENS

Waterside Productions

Cover design by:
Andrew Green
Books & Illustration

Printed in the United States of America

First Printing, 2020

ISBN-13: 978-1-951805-11-1 print edition
ISBN-13: 978-1-951805-12-8 ebook edition

Waterside Productions
2055 Oxford Ave
Cardiff, CA 92007
www.waterside.com

For you—
so you know for certain that you are the change and
you have the power

Contents

Introduction

Do Unto Earth is an extraordinary conversation intended to quantum leap us forward in our spiritual evolution and journey to enlightenment. This message is not a directive delivered from a thousand feet up; this is a very personal message from and dialogue with the Divine Wisdom Source directly to you and for you. Please accept this gift with eyes clear and wide and open.

Within these pages is the blueprint for environmental repair and peace and unity on Earth, however, this booklet constitutes just one of eight volumes that together make up that blueprint. While we believe that the eight topics, as separated by these volumes, are to be understood as connected to each other and only together give the full message as intended, we also understand some readers prefer to focus on their specific areas of interest—hence these eight mini-books by volumes. (Note: Chapters within this volume are numbered as they originally appeared in the book's full-length version.)

As you begin this journey, you might like to know how this collaboration of writing began.

It is indeed my great joy and honor to communicate with the Spirit Messenger, Pax, channeled by Carole Serene Borgens. From a young age, Carole, a former nurse, diligently studied all things metaphysical. This Spirit Messenger first visited her in the early 1990s when she was new to channeling by automatic writing. When her pen wrote the opening introduction and request for her to be a channel, she recognized the profound responsibility attached and jumped up from her office chair to pace the floor—not easy with three sleeping Irish Wolfhounds covering the carpet. Carole's initial response was to ask if she could think about it and take some time to respond, which she was given. Asking, "Why me?" Spirit responded to her: "You are new to this, you have no bad habits, and you will change none of my words." In time, Carole came to be comfortable with this blessing and so began her journey.

I, too, have been a seeker and spiritualist since my years as a teenaged runaway, and so it is a useful tool at times for me to reach out to a reputable intuitive for deeper guidance. Beginning on the fourth of February 2019, I had several long-distance Spirit channeling sessions with Carole—she was in British Columbia and I was in Florida. I had copious questions for Spirit as I sought further direction for my second title, *Do Unto Earth* (which, incidentally, is also the name of my business), while building upon the message of my first title, *The Magic of Viral Energy.* I was expanding and broadening the message of "viral energy" from personal and interpersonal goals to global concerns facing humanity and Planet Earth. I was also simultaneously establishing the Viral Energy Institute, a learning and research platform for the study of Viralenology.

Through our talks, this Spirit Messenger and I were getting to know each other and Spirit felt my passion for the plight of abused animals and species extinction, as well as my intention to bring awareness to our environmental crisis and to share the impacts of "viral energy masses"—large energetic fields created by both light and heavy intentions and action by communities, populations, industries, governments, and cultural beliefs—on Planet Earth. These disruptive energy masses create massive vibrational pockets of particular energies including love, hate, peace, discord, gratitude, violence, forgiveness, indifference, and compassion.

The Spirit Messenger seemed very interested in this direction and before long, Carole contacted me to say that Spirit wished to offer wisdom to be used by and shared through the Viral Energy Institute regarding this mission of planetary healing.

The writing began on the second of October 2019 when I sent questions to Carole who then channeled Spirit's responses by automatic writing (today, she does this via typing). It was *during* the writing that it became clear to all that this conversation would take book form and adopt the title *Do Unto Earth*.

As the answers were returned from Spirit, Carole and I both had many moments of excitement and more than a few gasps followed by, "Ooooh crikey, this is going to change everything!" The first of such revelations came in Chapter One when I asked the Spirit Messenger (whom self-identified with the moniker

"**Pax**", meaning peace) to be more specific about who they are. Here was the answer…

> "**We are one with the Universe, not the Universe alone. We are the Divine Universe, yes, and the God being and the greater wisdom, that which knows and supports all and is healing, non-judgmental and tolerant, all-seeing, all-knowing, and Peace.**"

Volume 7

Do Unto Earth Pax and the Next Evolutionary Leap

"I thought that thirty years of good science could address these problems—I was wrong. The top environmental problems are selfishness, greed, and apathy, and to deal with these we need a cultural and spiritual transformation. And we scientists don't know how to do that."

Gus Speth

Top U.S. advisor on climate change, former Dean of the Yale School of Forestry and Environmental Studies, and winner of the National Wildlife Federation's Resources Defense Award

Chapter Twenty-Two

What's Next in Clean Fuels

*W*e have traversed a lot of ground so to speak, and now let's turn our attention to future fuels. On this front, will you please help us to be the change we wish to see?

Yes.

Thank you.

Pax, we can no longer wait for scientists to make the progressive discoveries because in many cases (though not all) the scientists are funded by massive for-profit companies—often owned and operated by traditional fossil fuel interests.

Ground oil is neither sustainable nor clean and will indeed be replaced when greed and corruption within the fossil fuel industry is replaced with saner heads. Until that time there is behind-the-scenes exploration of alternatives, and technology is there but it is hidden from view in fear of replacement of the massive monies being lost by corporations in control.

Yes, and it's not fair to the masses.

May you please tell us which technologies to pursue that will replace the fuels and tech that we currently use and that is so damaging in terms of emissions and pollutants?

In past times and future times, parallel universes and off-planet technologies evolved to the point of thought power.

There is no need for fueled vehicles as teleportation is the thing; there is no need for burning fuel for heat as cracking together ions in technology produces what is needed.

The replacement for crude oil, in terms of propulsion and heating/cooling and more, is not a product but *a method* of alternative fueling. Some methods are in existence and in use, such as solar, wind, and other liquid-type fuels, gasses, propane, and more. They are and it is available and in use in some areas but not enough.

There is a large gap between where you are today and these methods, so an evolution takes place bringing electric power into use more and eventually giving over to a replacement with solar and wind and hydrogen and then to ion-specific power. Clean energy replaces all for the betterment of Planet Earth.

Moving ahead in time, your world will self-destruct at the current pace of pollution. It is for you to consider now that beginning change in transportation is crucial to the survival of your air quality. There is no way around this.

As you have created a population whose need for instant gratification extends to all things, you have also created the monster of gasoline engines overcoming your air to the extent it is brown in major areas. Think of this—brown—and you allow it to continue. What is wrong in this picture?

It is time to slow down and think of the alternatives.

You say there is a large gap between where we are and future clean energy sources. Let's look at it: Are we underutilizing solar power? Will we develop a completely different way to think of and use solar energy?

Solar power is almost a thing of the past while it tried to become the future. Technology moves rapidly for you now and this passive source remains a reality, a resource: it along with wind-power, in certain regions, can be a source of reversing current overuse of hydropower. There is more to come in energy sources and as your world turns, so do the options available. They will be revealed in time.

Rather than collecting solar energy in a passive way as we do now with the use of solar panels, is there a way to use the sun more like an interstellar charging station?

Within current technology there is experimentation and change comes. To harness the sun-power is seemingly now an old technology, but there is much new to be understood and it will as time goes on, yes. Filtering the heat and separating it from the burning rays is a portion of this development, and using the ultra-violet aspect for other purposes exists. What can be accomplished here is huge and more will be known in your soon time.

Gosh, this is when I wish that I was a physicist or engineer as I'd love to dive into this sentence that you give us regarding the future of how we actively use solar power: *"Filtering the heat and separating it from the burning rays is a portion of this development, and using the ultra-violet aspect for other purposes exists."*

Let's hope the baton of wisdom is passed to those who can work on development.

And now to the subject of fossil fuels and ethanol, and the move away from their use—this we say is the way.

To reduce or eliminate fossil fuels from your reality is a campaign for many now, and it is beyond time to manage this into reality.

There is no further need for dirty fuels. What the need is now extends to fuel to run the electricity used for non-combustion engine vehicles. Clean fuels from nature are useful in reducing to fuel sources.

Farmers will retool and change crops from those not serving them well to something of a future use on a grand scale.

Is hemp the crop that will serve them better, or something else also?

It is one, yes, and when fully utilized as replacements for your world's great uses now, the farmers will want to adapt to this as commercial uses will be great for products, not food specifically. Textiles, containers, food source and so much more can be achieved through using all components of this product. And it is rapidly growing and sustainable.

Is there a fuel source that we're testing that is a waste of our time?

To use soy for heating and propulsion has not been fully considered: that it will not sell on a world market must be thought of and to retain these crops for development at home is the way. More is to be discussed but again, looking to the past to find the future is the way.

To clarify, are you saying *not* to pursue soy for heating and engine propulsion? Some of the arguments against using soybeans as biomass is that it would heavily soot unless shelled and the oil extracted, and also that the high oil content

PENELOPE JEAN HAYES, CAROLE SERENE BORGENS

would cause heavy black smoke. And yet, soy is being considered as a heating bio-fuel.

It is the case that it cannot and should not be utilized for this purpose. There are clean fuel technologies: look ahead to these rather than attempting to retool and clean up after a source not well suited.

What about soy as a fuel for engine propulsion?

Not a source of propulsion to be spending time on developing when others are highly advanced, clean, and next-generation fuel sources of non-fuel propulsion.

These are known in your labs and are to be trusted to be revealed when the time is right and when the engines for use are ready. This applies to intergalactic travel specifically, and while your current jet fuels are efficient, there are alternatives coming that are cleaner and less destructive to the biosphere.

We wish to consider the new uses for current crops and say the farmers need not fear, but they do need to remain open to change. Technology as it develops now will affect their decisions and serve to guide them to a place of energy renewal and Earth stability.

The farmer, once again, becomes the producer and savior of the people, just like in past times.

It sounds like something exciting is coming soon.

Is there another heating and cooling power source worthy of some discussion that we haven't yet covered?

A form of gas that is non-toxic and clean and is filtered from outside air. We point to the availability of sustainable energy found within the Earth air and which can be extracted.

A gas in our regular air can be a heating and cooling power source? Let me think about this.

Hmmm, let's look at the components of air: 21% oxygen, 78% nitrogen, and the remaining 1% is a mix of argon, carbon dioxide, and methane. Well, wait a minute—our air also includes atmospheric humidity, otherwise known as water vapor. A water molecule has three atoms: two hydrogen atoms and one oxygen atom. So, therefore, hydrogen can be added to our list of gases available in plain air. Is hydrogen the gas in our air of which you speak? Is that the gas that you refer to for heating and more?

It is a component of what may be when extraction is accomplished, yes.

After some digging, I found something that seems to fit. Is the following method the one to which you refer in using hydrogen from the water vapor in our air? "Photocatalytic water splitting into hydrogen and oxygen using semiconductor catalysts is an effective method for converting solar energy or sunlight into clean and renewable hydrogen fuel. This process is the most promising and renewable choice for the generation of hydrogen."

It sounds good, yes? And it is and will be when focus is placed on development.

Okay, we now have an interesting method to look at for heating and cooling in particular. I want to be sure that I'm not missing anything else: There are also scientists working on hydrovoltaics, the molecular generation of electricity from water. They discovered a bacteria protein (Geobacter microbe) in air that links together and with the help of atmospheric moisture, the protein chain can generate an electrical current by way of a patented generator device called Airgen. However, the power generation output is low at this stage of development.

Is this technology worthy of attention, or is the extraction of hydrogen gas from atmospheric moisture the direction of which you speak?

You have not yet progressed to the understanding of combining techniques and ingredients to simplify the process. When all is said and done, it will be noted that at this time in history the operational method of power creation and the components required were with you but not seen by you.

You've said this twice now and I'm paying close attention.

Water vapor extracted from atmosphere is doable and may be the pathway *for now*. It has applications in some areas, and that would be helpful.

Non-fuel methods exist, such as magnetic, and are in development that will take them far above your water vapors and gasses, these will go the way of the dodo birds also and become a blip in history.

Chapter Twenty-Three

Future Travel

*W*hat is our next small step to look at with regard to our vehicle fuels?

Your Earth science is on it. The technology is there, development in process, and clean energy—when it is permitted to surface without challenge by old industry—will sweep the present into the future. There must not be interference for the sake of money—from "those"; they know who they are—and there needs to be inclusion of their modified industry into the future fuels movement.

It (your next small step) comes in the way of extraction of a useful substance from the current source, useful as an additive and support fuel for future transport. It is known now to science; it needs to be known to the people, also.

Do you know there is a correlation between hemp and ethanol found in gasoline fuels?

No, I don't know about that. I know that ethanol comes from corn. I know that decades ago Henry Ford created the Model T automobile to run on gasoline as well as hemp fuel, and that some researchers are again looking at and testing the efficacy of using hemp oil as an environmentally friendly bio-fuel to replace non-renewable fossil fuels.

We've been drilling and fracking for one hundred and fifty years and it is high time we fully replace crude oil. So, I'm going to straight-up ask you to impart us with future fuel ideas and tech. Please do tell: what's the scoop on the correlation between hemp and ethanol (which is basically just alcohol) that is found in gasoline fuels as an additive?

Yes, it is the case that hemp can be a component and an efficient one.

Let me state the downside of ethanol.

Currently, it is ethanol that is used as an additive in gasoline. It's made from corn in the United States, sugar in South America, palm oil in Southeast and East Asia, and rapeseed in Europe. For a minute, let's talk about corn. By far, the lion's share of the world's corn supply is being burned up in gasoline as the additive, ethanol. (The United States is the largest producer of ethanol in the world.) The ethanol industry burns up sixteen billion gallons of ethanol per year as a gasoline additive, and this increases every year. It takes twenty-six pounds of corn to make a gallon of ethanol. Holy corn! One acre of land makes about 328 gallons of ethanol per growing season.

Due to this insane market for ethanol and the incentive to profit from making ethanol, indigenous prairie lands have been converted into corn crops. This only serves to increase the greenhouse gas footprint as carbon that was formerly stored in the grass is released into the atmosphere (where it joins other pollutants like nitrogen and combusted gasoline used to process the corn crops). Not to mention, the vast expanses of corn fields are tilled at the expense of the local wildlife's habitat. All of this still says nothing of the argument that corn is a food grain and would be better used to feed the world and end famine.

Please jump-in.

Hemp is a replacement component for others in short supply, or the land on which to grow them on short supply or poisoned by pesticides and unclean air and water. There are rainforest and other location crops not previously used while some are gathering wealth in keeping their products in high demand.

It is not only about size of growing area but rather about efficiency of use as a component.

Ah ha! So, hemp in place of ethanol in our gasoline is actually more effective.

A change is afoot to revise thinking of how to use resources and to amend what was with what will be. We are saddened by the loss of clean air and water and soil in which to provide what is needed for your continuation.

Sure, it's sad. It's also asinine, what we do to our environment.

As the environment becomes a bigger topic and a topic of debate, some environmentally conscious individuals are grappling with the idea of no longer travelling by airplane. However, life today is such that taking weeks to cross the ocean by sailboat is not realistic for most. Should we stop our air travel?

Should you throw out the baby with the bath water? To stop airplane flight would not be possible in your society. Commerce would grind to a halt as a result and what do you foresee as replacement for airplane flight we ask?

It is for your science to create fuel sources that either do not pollute or do so to such a minimal degree it is known to be the very best you can do.

Until your people learn teleportation, transportation will be on wings and wheels with ships to fulfill water-crossing needs. Fossil fuels are not needed for propulsion of any of these—we speak of alternatives so take heed and work toward these ends.

Hemp has turned out to be a superstar plant as a plastic replacement and an additive to gasoline to replace ethanol. I wonder what else it's good for. Could hemp be developed as a fuel for propulsion of our airplanes as the alternative to fossil fuel?

This is but one of the potential replacements now being considered by your science, and it is one that will *not* be top choice in the end.

What will?

There are derivatives of gasses that stand a chance of being useful when modified, as also there are atomic or nuclear derivatives that could prove useful when modified sufficiently. Much is currently in place as is experimentation on not just replacement fuels but internal combustion engine design change to the extent that acceptance of alternative fuel sources is viable.

To reduce fuel weight in aircraft is a benefit as well, enabling larger payloads and cleaner operation.

Liquid fuel is not necessarily the way in future, it seems, even if held to by those reluctant to change.

What do you think of electric cars? The upside is not burning fossil fuels, but the downside is that electricity comes from methods that are not all that clean—we don't often think of this when it comes to electric cars.

The production of electricity requires fuels that are not clean in the overall picture of this means of transportation.

At the level of automobile use in your civilization, there is no way to reduce emissions either in the travel or the production of the method of propulsion. Until your way of living is altered and your lifestyles change to include less travel on a daily basis, the need for travel vehicles continues and increases. First the lifestyle changes and then the lesser need for transport. Cycling and walking are the way, of course, but in your current world it is scoffed at by those whose fast lifestyles require fast transport.

Always look to the basics for answers.

What *method* of alternative fueling can replace crude oil use altogether?

They are many and varied and range from *magnetics*, to *fission and fusion*, and on to what you consider *batteries*.

However, the future battery will be miniscule in size and via *magnetic fusion* will propel *transportation* methods of your future.

(Note to reader: magnetic confinement fusion generates thermonuclear fusion power that uses magnetic fields to confine fusion fuel in the form of a plasma.)

There are those currently experimenting with propulsion techniques that leave fossil fuel in their dust.

That's very good news.

When the clouds part and the way becomes clear, a burst of development becomes evident that takes your people higher and farther than you might now expect.

It is the fate of all to be busy in one place while developments come in another.

That's an interesting statement.

In your future time comes the organization of science into different departments and new ones emerge for study.

From what sector does the future method of propulsion come?

It is the section of physics that focuses on particle acceleration, in small scales but *greater intensity*, to be the place to look for propulsion and fuels of all kinds. Soon enough there will be breakthroughs in physics that will be declared monumental and ceiling shattering.

Look to your current students to know the secret and find the way, to show the way.

The development of an ability *to split what was thought to be solid* and divide with fission becomes a daily reality that changes the way of doing things.

Looking to that a little further shows that this ability exists and is being used in *other ways*. Look to it for a spark of change in the ways of heating and cooling, moving through space or moving through villages. There are options and small *masses* produce *monumental* amounts of energy.

And it brings clean to the forefront of human life on Earth. It cannot be monetized as fossil fuels but when time passes and those in the forefront of sustainable living have their power restored, it will be the focus of the future.

We leave it to you now to explore.

Do not consider this to be unreachable for it is yours to take and know that those other-worldly friends watch and support this breakthrough. It is to your benefit to find this means of great change as it takes your civilization to a new level. All is to be well with it.

This reads like instructions and can be used as exploratory steps to follow.

Very specifically, what is the best next direction for our travel needs (prior to the advent of using thought for teleportation)?

Particle acceleration is the thing; it is a thing and in active practice in science. To apply it here and now, in your time, is being done within science. It has yet to move into the daily reality, but it *is,* and it exists and can be harnessed by those capable.

May you please expand on this?

Particle acceleration is a scientific process that produces energy—this is a simplistic explanation. Like splitting the atom, it brings new and powerful sources of energy, specifically jet propulsion; an integral part of your Earth life, is aided.

What is the best use of this particle acceleration technology?

Should you desire to travel to the moon or the stars or other planets, fuel is needed. Until such time as teleportation is mastered, these processes will continue to develop.

An ion thruster is the current form of spacecraft propulsion, however, at present it doesn't get us to other planets at light-speed or nearly fast enough to make other-planet travel reachable. What can you say about the mention of "ion-specific power" in terms of how it is different from the ion thruster?

We spoke of acceleration and particle acceleration and transportation. Use of these technologies extends well beyond "fuel".

If the particle acceleration that you speak of is "*like splitting the atom*", then perhaps it's *not* the atom that you are suggesting accelerating. You also spoke of "cracking together ions" for "ion-specific power". Is the future-fuel mentioned for use for on-planet travel and for heat and fuel needs the very same as the propulsion energy mentioned for use in interstellar travel, or are these two different technologies?

The ability to utilize existing fuels for heat and land-based transport remains constant. The use of liquid fuels for off-planet travel will cease as an alternate method of propulsion beyond light-speed is further developed. It exists in theory and is becoming reality in testing by your scientific community.

When you speak of particle acceleration, is this the same as what scientists are researching at CERN, the European Organization for Nuclear Research?

We spoke of the evolution of fuel sources that rapidly now become propulsion sources utilizing no fuel. It is beyond this current discussion.

Okay, not CERN.
"Propulsion sources utilizing no fuel" is intriguing and I'd love to explore this. Will we get to the crux of it if I ask the right questions?

Your questions are quite acceptable. It is the out-come of exploration that shall take further time and we do report on what proves best for your uses at a *later date.*

While the reader will flow from your last comment to this one, today is a later date. To be fair, it's only days after our last words on this topic, but I do have another question to flesh out.

Last night, at a dinner party, someone men-tioned that the big answer to future interstel-lar travel will be "*plasma* propulsion", and so I wanted to ask you about it.

I found this description on Wikipedia for two types of propulsion modalities, each with their pros and cons of efficacy: "A plasma propulsion engine is a type of electric propulsion that gener-ates thrust from a quasi-neutral plasma. This is in contrast to ion thruster engines which generate thrust through extracting an ion current from a plasma source, which is then accelerated to high velocities using grids/anodes."

Both sound like they could be a lead on the no-fuel propulsion systems and/or "ion-specific" power source that you hinted at. I don't know if I should bother you again on this topic at this time, and yet when someone at a dinner party brings up interstellar travel out of the blue, I wonder if this is a fluke or how the Universe whispers to me. I'd love for this work that we're doing to help our scientists so that they might then focus on

the tech that's best for the purpose, and so my question is: is one of these more on track than the other in terms of the future fuel or future travel that you had referenced?

It is the case that ion-thruster is the work of the present toward the path to the future.

As it becomes easier to influence the ways of science to invest in this mode, it will grow further into commonality. Science trials repeat and grow and change and return ideas and theories and postulations, but the "meat" of it is in the doing, and we say those currently practicing with this method—testing and experimenting with it—are close to stating this will do the job.

Moving through space and time will become a need and currently is a want, so those involved are feeling pressured to develop a working model. We say they exist and are being tested in reduced size and power.

For these future travel technologies to not be hoarded by big industry or hidden away from public access, it would be helpful to know what exactly we are talking about so that the people can be part of the process. I'm enjoying putting this together, yet I want to make sure that I'm following the breadcrumbs down the right path.

This is the time to know that future technologies are presently in scientific laboratories and in testing stages. There is knowledge in select segments of science, not general knowledge, not ready for publication outside of science, but working knowledge and a believable piece of, perhaps not-so-distant, propulsion to move people and cargo through space. This hasn't a name at this time nor a pathway to existence that we are ready to examine, but suffice it to say, it will be. The end result is that planetary travel will be managed with the ease and fluidity as that of your intergalactic visitors.

Who are our intergalactic visitors of more recent times?

Friends from other galaxies. As we have said, they continue to "buzz" by to see how Earth peoples are faring but see nothing they want so continue on their way. As they believe Earth's peoples have decimated the planetary resources, they are saddened and don't wish to remain there.

Are there other peoples on other planets in the Universe that need this much babysitting from the Spirit World and buzz-by monitoring from intergalactic supporters?

There are and have been, yes, and they come and go as lessons are learned.

Many people have wondered about our technological leap forward in a relatively short period

of time. For instance, our computer age seemed to arrive all at once, and it was a rapid quantum leap. Did our computer and Internet technology come from Spirit, from interstellar visitors, or from our very ancient starseeded ancestors?

Largely responsible for this quantum leap were those starseeds who brought their personal ingenuity and advanced capabilities to the technology: these young leaders in the science propelled development and still do.

Young leaders?

Young leaders in the science—this means these leaders are *young in the science*. The science is also young.

As there is nothing new under the sun, your technology comes from the past to fuel your future. Your inventors who have come from past lifetimes in advanced civilizations are the disseminators of information that now moves you ahead in the practice of technology shifting. It is a time of greater movement in this sector and brings excitement to the fore as it pertains to inter-stellar travel development.

Will I be alive when interstellar travel is commonplace?

We suggest that interstellar travel is reality now, in your time, and will extend to widely practiced means of people movement within your generation. It is not now considered a regular means of travel for public use but will be. As the need on Planet Earth escalates and as the movement of off-planet exploration swells, the need will be met.

Your hydrogen and other powered transports are being finessed into another form that can journey light-years away in a speed previously unheard of. It is on the cusp of discovery now and will enable the thought to become action. This will determine the when, further science the where. In our speak we say the far time which, for an undertaking of this magnitude, is soon.

Can you tell us something—a clue—about what we need to look at to make this discovery?

This is a present theory for interstellar transport using no fuel, as you know it. Currently it is being made known to your scientific community and has far to go in development, but we say it will become reality and change the way of off-planet transportation. It provides for light-speed which enables otherworldly visitations routinely.

It is "being made known"…hmmm, interesting choice of words. Are our scientists creating no-fuel propulsion technologies from their own minds and through trial and error? Or, are they being *given* the information, such that the technology is not their own and is instead from

otherworldly beings or even from Spirit-inspired thoughts? And, furthermore, are such discoveries given only at a time when we can "handle it" and use it for good?

Oh yes, we say that when the student is ready the teacher appears, yes?

The no-fuel fuel, meaning no liquid fuel. Was it a channeled inspiration? Quite possibly. Where do brilliant ideas originate—from within or not? What is from within and do experiments take their course when one step leads to the next and a brilliant mind has an "aha" moment? Some of both, we say, that brilliance comes through experimentation, knowledge and divine inspiration also. It's a process and remaining open to all thoughts and ideas is key.

For the currently unknown no-fuel power source, is the following correct: *hydrogen finessed into another form created by smashing together ions is the "fuel" to be used for interstellar transportation.* **Am I on the right trail here? You have said that it still has "far to go in development." What is my next breadcrumb on this? Or, if I've incorrectly put this together, please correct me and show me where I'm off.**

The no-fuel solution to interstellar travel is in development now and will bring an exponential leap forward in space travel.

(I might be getting a parental pat on the head here.)

While I'm straining for an answer or a formula, I'm starting to get the idea that you're holding back for a reason—likely a good reason. And, it could also be that I'm not putting the pieces of the puzzle together very well at all.

It is known and will become a race to develop. With North American science and European science collaboration, the best will come in the way of discovery. Funding, of course, is always a deterrent when sufficient isn't available. As this project is of the highest priority, for some, it will soon become widely known and funding will be directed to it. Earth is not ready for this, but it will be.

Space travel to this extent will be fought over by private enterprise and become a scramble for superiority. Again, your people show their dark side. Nevertheless, science moves forward and those in the know now prepare for ways to utilize this discovery for their own advancement. It is a sad place your people have come to. We wish to scold those who know better and can use their influence to ensure idealistic values are retained and practiced in your civilization.

Carefully walking on thin-ice and with great respect and gratitude, I will go at this again. Speaking of the no-fuel power source that could power vehicles in space and on-planet—plus

replace electricity, oil, coal, nuclear and all polluting and/or non-renewable energy sources—if
this future power source is only in the hands of
the few at the top and is therefore patented and
monetized, the rich and influential are the ones
who benefit and they become richer and more
powerful, which further engenders disempower-
ment and poverty for the many.

That being said, I feel the responsibility to
advocate for our people. At this time, what more,
if anything, can you say about the no-fuel power
solution?

Ah yes, this solution is in your labs now with
the scientific community developing further test
methods.

The particle acceleration aspect of this clean fuel
brings your world closer to the edge of space on a
regular basis. There are *more uses*, of course, which
are revealed in your soon time.

For now, we say that the discoveries of today
prove to be the basis for tomorrow's transportation
methods and the advancement brings rapid change to
your manufacturers—a quantum leap forward for all.

Like teleportation in a sense, it mimics the think
and move through time and space way of going. The
no-fuel aspect of this is to remain just that, for now,
as there is no fuel needed and *soon all is revealed*.

You are very patient with me and I appreci-
ate that. I am (finally) getting the message loud

and clear, and that is that you are holding off— at this time—on giving us all of the answers to future fuels for interstellar travel, likely because you believe that the information could get into the wrong hands. Is this correct?

It is our pleasure to speak more on this topic when the time is right. For now, there are lessons to be learned by your Earth people to satisfy the need for cleaner living and cleaner energy and cleaner intention before the Universe provides and releases the knowledge to your science that enables such travel.

The planets with life-giving environments are there but inaccessible to you at this time. Undisturbed and clean they support life and civilization and await visitors. The sadness here is that the minute in time your Earth people do arrive they begin polluting and disturbing and leaving litter and signs of themselves—in similar ways they have decimated Planet Earth. Your "space junk" and debris left behind on your moon is an example. What is left behind is done, but going forward there must be concern for another environment's wellness, not just how to get there and claim it. This will not do. Until this practice stops there will be no forward motion in this ability to deposit yourself on the soil of another planet.

This is the one thing to be aware of: The Universe provides in many ways and protection against flag-planting and claiming of territory and ability to have no boundaries will not be tolerated. Only when your people become kind and aware of their responsibility in this matter will the ability to find alternatives become reality.

To the extent that your people consciously determine they will function with zero waste debris, zero plastics, and heightened respect for another planetary home, success in finding the optimum alternate location will be yours. Until then, there is much attitude adjustment to be undertaken.

To learn propulsion for space travel is one thing, but until your daily damaging of the environment is ceased, no amount of balance exists. Your obvious choices for the use of non-recyclable products to support life is a sadness to us and the obvious alternatives exist and are not recognized by industry as sufficiently profitable therefore are not given the research and development budgets needed to become mainstream.

Again, we speak of corruption and greed among your politics and your industry and commercial ways of manufacture. Too much is known to be abandoned as reality. This is detrimental to your future in all ways. We are displeased by this bold defiance of what can be healing for your planet.

We ask that the people of Mother Earth clean up their own planet and learn new ways of daily life. We are here to say that the future as you foresee it will be and the ways of travel and accessing a new planetary home will become known. Until then, the focus is to be elsewhere: on the corruption and greed to your government and your industrial giants and your corporate executive need for control and wealth and none of this qualifies your civilization to deserve a clean and pristine new home.

As the Universe watches over your actions and condemns them, the Universe also blocks vital technology from being yours if it is designed to overtake and damage another planet.

Your people are shameful in their actions toward your Mother Earth and are not welcomed elsewhere in your Universe. Harsh as this sounds, it is designed to cause soul-searching and inner soul visiting for each person's ability and desire to clean up their personal act as well as move to clean up for the greater good.

Chapter Twenty-Four

The New Scientist and the Seeker

Pax, I wonder what's the trajectory of our evolution and what will quantum leap us into action for the greater good.

You once mentioned "the next age of your people". I didn't follow it up then, yet I'd like to do so now. What's the next age?

Moving ahead in time, we see generations of self-motivated people working for the cause. This is needed. The cause is the preservation of your environment and your world as you know it. There is *a new breed of scientist coming* which revolutionizes the way earthlings operate and this will become evident in the soon time.

Is the new breed of scientists to be the Viralenologists that I've envisioned through the work of the Viral Energy Institute (*after all, this*

is what brought us together for our writings)—the new scientists whose work includes the use of out-of-the-box methodologies of spirituality and metaphysics?

Yes.

Talk about coming full circle!

In that case, I would like to enter into our record here this new term and occupation that I feel should be commonplace going forward. Here's the definition:

Viralenologist (noun)
1. One who studies viral energy in social interaction, interpersonal communication, and for the betterment of individual relationships and personal development
2. A scientist or intuitive who studies Viralenology in environments and as the causation of large planetary pockets or masses of enlightened or heavy/dense energy
3. One who studies Viralenology for new and renewable energy sources, future modes of travel, new technology, and communication
4. An intuitive who accesses higher-stream-consciousness (a.k.a. Universal Intelligence) by way of the viral and osmotic nature of energy in order to learn from and share the wisdom-of-the-ages

Your goal, Penelope, can be to inform and educate now so your people change their thinking and acting and do not, in future generations, continue to do damage in other worlds. What you teach now will be taught down the line; future generations benefit, and your legacy continues.

As you know, I'm a believer in spirit messengers, conduits of Universal wisdom, empaths and those who intuit energy fields, channelers of Spirit, and those who have tuned into Higher Self wisdom through practice and intention. I believe that we need to tap into their line to higher consciousness and have them teach us the ways in which we can develop our own higher abilities. I also believe that we need to elevate Spirit channelers to important roles in our scientific community. If scientists would pair up with good and reliable channelers of the Spirit World, we would more quickly evolve in our technology, consciousness, and forward movement in all ways. Do you agree?

Yes.

I see a future where intuitives are regularly consulted by our scientists of all kinds, from astrophysicists to the engineers of our future fuels, and from medical doctors to mental health professionals.

There is indeed infinite wisdom to be mined when there is a meeting of the minds, as it were, and the resource is there for the taking. The great distance between science and metaphysics is understandable, between traditional organized religion and metaphysics also. There is doubt of value from one and fear from the other. Neither will mellow in the soon time, but Penelope, your stating this need for connection will begin a shift.

These could include the metaphysical practitioners, energy healers and teachers, researchers of the contagious nature of energy, students of the osmotic properties of energy and energy sharing, intuitives, mediums, and light workers who utilize spiritual practices and sixth sense abilities within science and traditional studies of science to enlighten themselves and the world.

This is the way of it beginning now and going forward.

There is much to be shared with those who are not on this higher plane of exploration of sciences, and it will take time to unfold. This is as it should be as during this phase many more of your Earth people are learning and understanding that they can make a difference and will be a part of the solution, not a part of the problem. And this is a lesson they will teach daily as they go forward into the population around them.

They are the New Orions who bring light into the darkness—they are the stars shining light onto and into the darkness to illuminate what is there.

Once identified, they may begin to remove and repair and rebuild your society.

Is there any connection between your reference of the "New Orions who bring light" and the star constellation Orion known for its stars of bright light that comprise Orion's Belt? The three brightest stars are called the Three Kings and are known to be beacons in our night sky.

It is a name only, the name of Orion which indicates the brightest of bright.

These are the people of your world today who have seen the light and begin to bring their light into the darkness of unawareness among the population. The light is directed toward clean Earth resources and pure intention among the masses. It is the way of bringing about a return to simpler ways and purity of intention, honesty in all matters and a knowing that Planet Earth needs coddling and care, protection and respect, and teaching how this is done.

These are adults who have found their way to this place of understanding and found their need to teach and inform and share. These are the New Orions— those whose light shines brightly into the darkness of your Earth's present.

Wonderful. So, the New Orions *are* the Viralenologists—metaphysical seekers using enlightened thinking and feeling to arrive at

scientific discoveries and future technologies of all kinds.

Yes.

You mention that they are adults. Are they exclusively among the adults of today, or do they perhaps include some of the enlightened young people who will also develop into the New Orions/Viralenologists of the future when they grow up?

They are enlightened ones, yes, and not of a particular generation.

There are many historical references to Orion, and also semi-recent popular culture references hinting at deeper significance such as in the film *Men in Black* (1997) where the galaxy is found to be hidden in a charm on the collar of an ET's cat named Orion.
Is "Orion" a clue to something bigger for us to explore?

Yes.

These, the new breed, have in common the greater good and the big picture for your Earth planet *as well as your galaxy.*

Cohesive planning and action is called for by these individuals and groups—they have a higher calling in their adult years and have seen the way. Having a voice and using it for the well-being of Mother Earth is their intention. Together with those younger who are committed to the same cause, they take action and enable those who watch and listen to know they may also participate and create change where needed. They are clarions for the cause. Your world is blessed now with the uprising of minds and spirits and souls and bodies for world repair and wellness.

To recap, the New Orions are the new breed of scientists: enlightened adult-souls within all generations who have a role for the greater good and big picture of Planet Earth and the Milky Way galaxy.

Yes. They wish to bring their zeal and excitement for change-making to the forefront and allow their inner beings and inner-voices to bring their strengths to the project of saving the planet.

They lead and they learn, and they teach, and they are *inclusive of all who can contribute*. It is their joy to become one with the movement to peace on Earth.

What else should we know about them?

These New Orions are to be recognized and trusted, not shut down for unusual or creative ideas that fly in the face of tradition. Nothing new comes but repeating old ways, certainly in the repair of your planetary resources.

Now it is time to trust in their recommendations and be the voice that agrees to adapt to the new ways.

Within the movement will be those who *follow* and work to create change, as well as those who *lead* and work to create change—both are needed.

Quietly going where none have gone before in the intention to create wellness out of destruction: this is the mindset of those who see themselves as Earth healers, and this is the role now for many to consider adopting.

If the goals are to clean air and water and soil and the minds of those who pollute, all these are needing individuals who feel the call to make the change needed now by Planet Earth.

There will be political change within your view in Earth's soon time, and that will enable a more peaceful change to come. Those in power who monetize their positions will find themselves on the outside looking in, soon enough, and those with inherent honesty and pure intention will be held to their ideals and lead accordingly. Change comes and your Earth is the better for it.

Your people are to know that it is a long-term plan to bring about the repair and healing your planet requires, but with the purity of intention you will find in new leaders, this task becomes a joy and like ripples on the pond, this feeling of pure intention spreads throughout your globe and is felt and emulated by Earth people. It is a good thing.

I think we're changing. I feel a groundswell of pure intention for Planet Earth. Environmental concerns are all the talk, and some action, too. Individuals are becoming aware and making choices to reflect this intention for planetary wellness. It's true that our governments are still moving full steam ahead in the wrong direction for the environment, however, the people can (and perhaps will) be the change needed.

Pax, I'd like to think that we are empowered about the physical reality of our world; empowered to make happen the changes that we want to see. Are you making a connection between our empowerment and the ability of the average person to make the discoveries needed?

Yes.

It is a time of change and becoming one of enlightenment tempered with wisdom and humility for those who thought they had the answers—they did not. And for those who thought they showed the way, well they did that all right, but it was not necessarily the best way.

We speak of the forces of nature and that means what is meant to be in your world at this time, what is intended, and what will result in the trajectory change that is needed.

I'm unclear (and I think this is important): what are the forces of nature?

We suggest that a "force of nature" is a movement that cannot be stopped, like a hurricane cannot be stopped as it is both a force and of nature. Those who come your way now to be the change that is needed are the forces of nature.

This is compelling and exciting.

As we speak of the "peaceful ones who show the way", they are the starseeded ones with integrity and passion that *come from elsewhere* to you as a gift at this time. They are the Crystal and Indigo children, and also, they are the New Orions—we call them by many names and the names themselves are not of matter. It is their intention and their goals that are of import to you and your Planet Earth. These are the voices to be heard, the integrity to be felt, the intention to be followed and the harbingers of a new normal for your world as time plays out.

It is for your people to trust in this, watch and learn, and choose your channel for peace. Each has a role to play, each has a need to contribute, and this is the time to know that the need will match the giving and the givers. To that end there will be change shown in a time and in a way you would not have considered. It is your time and their time and when combined, only good will come.

Trust in this and trust in yourselves.

Chapter Twenty-Five

Communication Through Consciousness

*I*s the Universe one ginormous organism?

All is one and one is all and interconnected in this Universe. What is in one place affects what is in another. It is the way of it. It is the linking energy that holds the Universe together, as that Universal energy holds your humanity in its protective grasp.

Is humankind sort of a metaphorical body part or organ of the Universe? Do we have an impact on the scale of the infinite cosmos?

Interconnectedness relates here and one affects all as all affects one. As it is important to not have extinction in the animal world it is important to not have extinction of a planet's race of people. One feeds another and one cares for another and to have a missing link is not productive—there is a resulting weakness of the whole.

How would the Universe be weakened if humanity became extinct?

As with links in a chain, with one missing the strength is compromised or lost. As with the Universal energy, such is the case that all of the whole makes up the strength of the whole.

Gaps in space are there for a reason, a distancing, and intentional. For something present to become not present is a weakening. This applies to the food-chain and the brain-chain in scientific development, and it applies to all aspects of life as you know it. Think on it from a familial standpoint: a family is strongest with all members present; remove one and there is a void that is never filled. It creates a weakness of emotion and, depending on the familial activity, a weakness of productivity perhaps. Expand that to the Universe and it is clear.

According to some of our cosmologists and quantum physicists, the Universe is *physically* connected, and not just metaphorically, meta-physically, or spiritually. Research from these cosmologists has shown that all planets and all stuff (and non-stuff, for that matter) in a galaxy are connected and "held together" in balance. Furthermore, all galaxies are connected to each other, so they say. Also, super-clusters (groups of

galaxies) are connected to each other as if in a web. So, the whole Universe is connected.

Here are some questions based on these discoveries—

What is the mysterious power that connects everything? Is it a gas, an energy, or a force?

Well, it is the gravitational pull that is responsible for suspension of orbs in space. It is a push-pull result of being held together in the desired pattern. There is magnetic management of the distancing also, and planets and stars hanging in space do so with the magical threads of compression in place. Push and pull results in holding in place and not falling through the space they inhabit into another place. You do note that regularly there are comets and other pieces of the puzzle that break free and go streaking through the maze of orbs on their way to burning out. What is not intended to remain in space doesn't.

All is managed and controlled and finitely tuned to hold what is and what will be. Into the galaxies go the magical threads of what results in the dangling of orbs in the night sky for your viewing pleasure.

That was a beautiful description. You are a poet, Pax.

Speaking now again of our friendly intergalactic supporters, you said, "We access their wisdom through our conversations." I'd like to visualize how your communication with interstellar

supporters works whereby they may pass on their wisdom to you and then to us?

As we have a teleporting and telepathic communication model, it is our way to communicate and to be.

The visitors who have been and continue to support Earth people's development do so as that is their role. They give of their technology and they give of their knowledge generally and offer to show the way to higher learning and functioning. They do this throughout their travels in the galaxy and farther afield. Where they are accepted, they help to advance civilizations and society. Often there comes an end to that acceptance so they continue to watch and await a potential need for their future interventions.

Should you wish to access the wisdom of extra-terrestrials, there are methods which do not involve us. Consider the application of remote viewing and the Higher Self accessing—always there are times and methods useful or not (and the time is not now to consider a direct communication with extra-terrestrials): this refers to accessing our own Higher Self wisdom for the purpose of intuiting ET wisdom. Here are two separate, or three in reality, existences: theirs, yours, and ours—and all on different planes, alternate realities and methods of communication. Allow time for them to meld.

Tuning in to another place to see what and who is there and what are the current activities. This technique has been commonly used by government security of highest level in times of war and cold war, to infiltrate from afar the operations of those they need to know. Spying it can be called. Viewing through the training to do so and utilizing one's psychic abilities to see the scene and report the details. Drawing what is seen in the "mind's eye" is a way to transmit the sight to paper for record keeping.

It's kind of scary (and eye-opening) that our governments have been using this ability to spy. I wasn't expecting to hear that governments use (what I consider) spiritual practices to spy for their own gain. Shouldn't this be illegal?

You overreact here as this technique is mind power in use for good. To know what a national "enemy" is doing is practiced globally and not a new technique. You might consider this being practiced for the good of the country and its national security. This is a widely practiced technique that is not only not illegal but is fair play: all is fair in love and war it is said. The use of this practice is more acceptable to many than entering a battlefield and shooting to kill. In fact, this technique and the knowledge it brings can prevent such action.

This sounds reasonable.

Recreationally it is useful to view a target of travel or family or road conditions—it is open to possibility. The practice of this technique is that, *practice*, and identifying the place one wishes to see and focusing on it, allowing the Higher Self abilities and strengths to function to the extent that a mental picture forms. Some do this as recreation. Others have done and currently do for national security reasons. It is not rocket science, but it is for the trained mind to achieve and brings great value to security needs.

Okay, I'm satisfied with that. Thank you.

This next question is going to be "out there" (ha! pun intended), although everything that we talk about is rather out there, so here it is: I know that you've said that our friendly interstellar supporters (who used to be visitors) do not visit at this time because we are too hostile and our Earth people and political culture is currently too volatile. Sooo, my question is, can I visit them? A little off-planet rendezvous perhaps? Maybe they could pick me up for a day trip? I'm being funny, yes, but I'm also perfectly serious in that I would be up for the challenge and would invite a meeting of the minds if they're interested.

Until you are able to teleport, we expect this remains a daydream. Alternatively, Penelope, you may consider communicating through meditation: asking their assistance in understanding what more you can do to meet your challenges at this time is a beginning. Your intentions are honorable, and it will be known this way.

That you are an Ambassador for Peace on Earth in all ways is also *known*, therefore you may find contact.

I'm honored and eager to communicate via meditation.

Pax, these ETs are very interesting to us Earthlings; we're endlessly fascinated by them, actually.

I recall that you've previously said that we will be grateful for your guidance in this colonizing journey (yes! we are) and that you, Pax, "together with others of high-mind and experience in the Spirit realm, will continue to advance our support and ideas".

Who are the others of high-minds in the Spirit realm?

Those who have gone before and continue to observe and guide and send positive energy for the best possible outcomes for your Earth people are your teachers.

Their influence is felt in consciousness by those tuned-in to their Higher Powers and listening and following their hearts. There, in heart and mind the messages are felt and heard, the direction known as the right direction, shared for guidance and the knowingness that these are the ways now for your Earth salvation and healing.

How do we communicate with those guides in the Spirit realm?

Listen and learn, it is said, as above. Take the questions to your Spirit Guides, go within for answers always: this is the source for you.

I understand that ETs are physical beings in form. However, the Spirit World is just that—spirit. Through our conversations, I feel as though I know you in a way. Truth be told, I think of you as a friend and I even miss you between our chats. Is there such an experience as a visitation with the Spirit realm? To "meet" those of high-mind in the Spirit realm—is this possible? Can I *know* you in a way that is experiential?

Experiential indicates having an experience as a result, yes? So, your experience *is* in communicating with us and this is the way of it. We could, however, develop a fictional method where you might say, "Guess who's coming to dinner?" but this would not become reality at this time or in your future, as we believe you would ask.

Yes, I would have asked.

Be aware of this three-way conversation (Pax/Carole Serene/Penelope) being unusual by any of your standards and quite unlike the day-to-day experiences of most.

$\bullet \ \bullet \bullet \ \infty \ \bullet \bullet \ \bullet$

About the Author and Channeler

*P*enelope Jean Hayes is a new consciousness author, television personality, and speaker. She has appeared on-camera hundreds of times as an expert guest on programs including *Dr. Phil*, *ABC News*, as well as international news specials and telecasts. She is the foremost leader in the field of contagious and osmotic energy known as Viralenology, founder of the Viral Energy Institute, and author of the book *The Magic of Viral Energy: An Ancient Key to Happiness, Empowerment, and Purpose.*

Carole Serene Borgens channels Pax, the Divine Wisdom Source. Carole is a former nurse and longtime student of metaphysics. She has been channeling Spirit since the early 1990s when she was chosen by Pax and given the title "Spirit Messenger". Carole continues to write and provide in-person and remote sessions for clients around the globe, and she refers to her gift of channeling as "the greatest blessing in my life."

Of this trio, Pax says, "A good team we three."

www.PaxWisdom.com
www.PenelopeJeanHayes.com
www.CaroleSereneBorgens.com

www.ingramcontent.com/pod-product-compliance
Lightning Source LLC
Chambersburg PA
CBHW032017190326
41520CB00007B/515